DI TORIE A NUMBAS

THE NUMBER STORY

SMALL BOOK ONE

ENGLISH – JAMAICAN CREOLE

*Numbers Teach Children
Their Number Names*

written and illustrated by

MISS ANNA

Early Reader Edition of *The Number Story 1*
Bronze Medal Winner, 2016 Wishing Shelf Book Award

Library of Congress Control Number: 2018902040

Names: Miss Anna, author.
Title: Number story : numbers teach children their number names / Miss Anna.
Description: Portland, OR: Lumpy Publishing, 2018.
Identifiers: ISBN 978-1-945977-96-1 | LCCN 2018902040
Summary: The pictures and rhymes present stories which introduce numbers 0-10.
Subjects: LCSH Numeration—English—Jamaican Creole--Pictorial works--Juvenile literature. | BISAC JUVENILE NONFICTION /
Languages: English—Jamaican Creole
Classification: LCC QA141.3 .M57 2018 | DDC 513—dc23

Publisher: Lumpy Publishing
Website: www.missannabooks.com
Email: missanna@missannabooks.com

Paperback: ISBN 978-1-945977-96-1
Printed in the U.S.A.   1 3 5 7 9 10 8 6 4 2

Waan fi learn di name
dem fi we numbas?

It is very easy and a lot of fun!

It eazi eazi n ole eap a fun!

Say-along our little jingle

Sing wid wi like torie!

starting from Number One!

Wi ago begin a numba one!

1

ONE looks like my one finger.

ONE luk like me one finga.

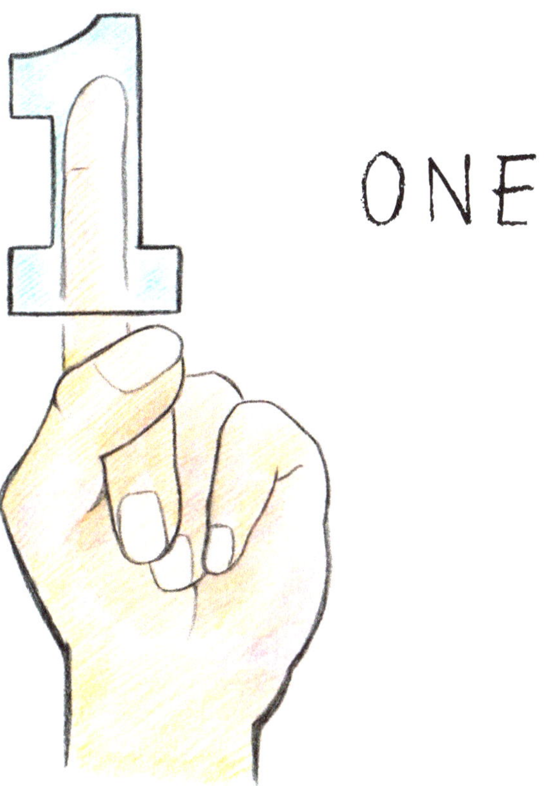

ONE!

2

TWO trails a tail.

TWO ave a tail.

A TAIL!

3

THREE has bumps.

TREE ave bump.

LUK PON DI BUMP!

4

FOUR carries a sail.

FOUR carry wa sail.

WAN SAIL!

5

FIVE is a racing track.

FIVE a wan racin track.

VROOM
VROOM
VROOOM!
1

6

SIX curves like a snail.

SIX curve lik snail.

WAN SNAIL!

7

SEVEN has a sharp angle.

SEVIN ave a shawp angle.

TEK TIME! IT SHAWP!

8

EIGHT is rollercoaster rails.

EIGHT issa rolla coasta rail.

RAE!

9

NINE is a bubble on a stick.

NINE is a bubble on a stick.

A BUBBLE!

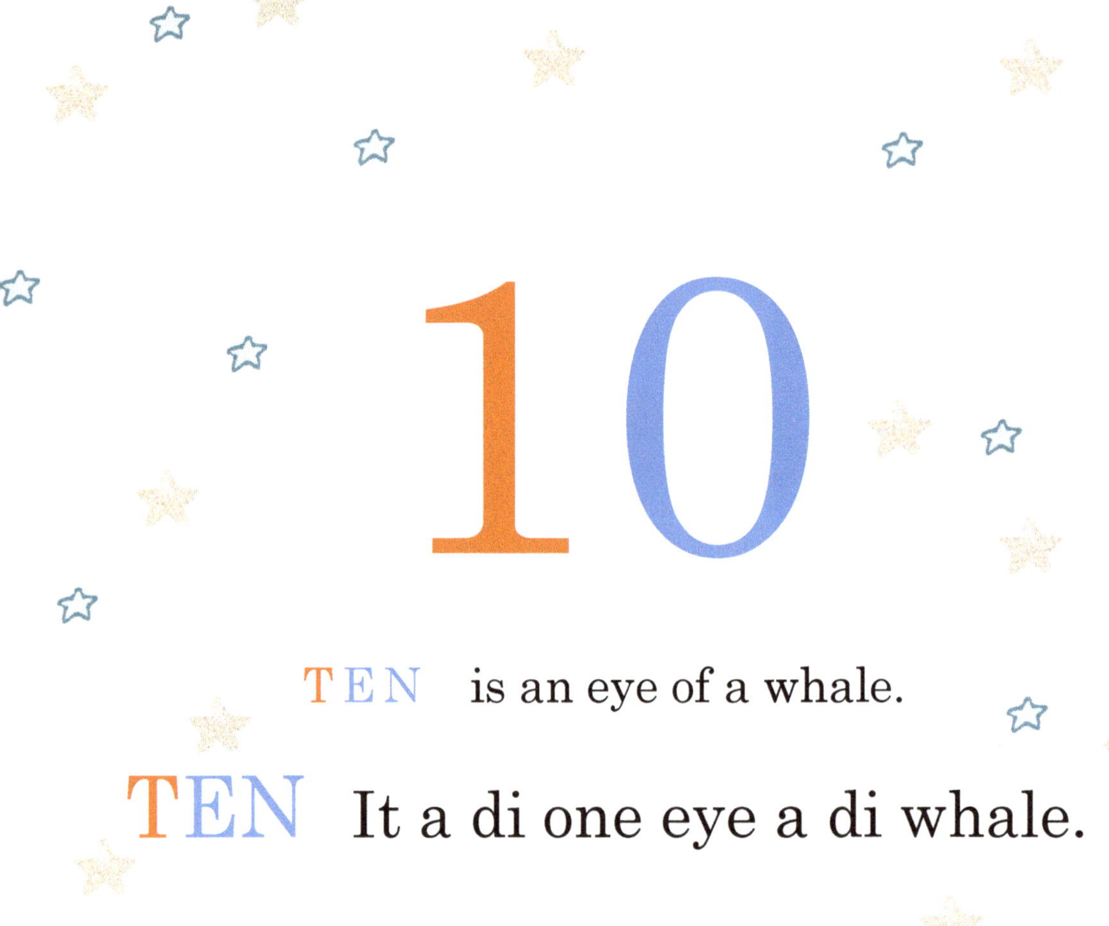
10
TEN   is an eye of a whale.
TEN It a di one eye a di whale.

ELLO!

And
An
0
ZERO is an empty pail.
ZERO issa empty pail.

IT EMPTY!

Thank you for playing with us today.

We had a lot of fun too!
Tank yu fi play wid wi todah.

Wi ave nuff fun too eno!

We are your Number friends,
Zero to Ten,
Who will be here for you~
Wi a yu numba fren
Zero to Ten.
We ago always deh ya fi yu!

Bye-bye now!
See you again soon!
Ba-bye!
Lata!